THE WATER CYCLE

Izzi Howell

Published in Great Britain in 2018 by Wayland

ISBN: 978 1 5263 0365 3
10 9 8 7 6 5 4 3 2 1

Wayland
An imprint of Hachette Children's Group
Part of Hodder & Stoughton
Carmelite House
50 Victoria Embankment
London EC4Y 0DZ

An Hachette UK Company
www.hachette.co.uk
www.hachettechildrens.co.uk

A catalogue for this title is available from the British Library
Printed and bound in China

Produced for Wayland by
White-Thomson Publishing Ltd
www.wtpub.co.uk
Editor: Izzi Howell
Design: Clare Nicholas
Fact Cat illustrations: Shutterstock/Julien Troneur
Consultant: Karina Philip

Picture and illustration credits:
iStock: ajman33 4, Hayri Er 5t, temmuz can arsiray 6c, 22kay22 13, Borut Trdina 16; Shutterstock: Andrey Yurlov cover, straga title page and 18, M. Pellinni 5b, Mila Dubas 6l, kazoka 6r, Henri Vandelanotte 7, Merkushev Vasiliy 8-9, Creative Travel Projects 10, Mike Flippo 11, Triff 12, Patrick Foto 14, jordache 15, Tom Grundy 17, paul Prescott 19, Martchan 20, john michael evan potter 21.
Every effort has been made to clear copyright. Should there be any inadvertent omission, please apply to the publisher for rectification.

The author, Izzi Howell, is a writer and editor specialising in children's educational publishing.

The consultant, Karina Philip, is a teacher and a primary literacy consultant with an MA in creative writing.

CONTENTS

What is water?.................4
Solids, liquids and gases........6
The water cycle.................8
Evaporation..................... 10
Condensation 12
Rain and snow................. 14
Underground water........... 16
Rivers 18
Controlling water 20
Quiz.............................. 22
Glossary......................... 23
Index............................. 24
Answers......................... 24

WHAT IS WATER?

Water is very important for **living things**. Animals and plants need water to **survive**.

Animals, such as these zebras, drink **fresh water** from ponds and rivers.

Most of the Earth's **surface** is covered in water. Nearly all of this water is **salt water**, found in seas and oceans. A small amount of it is fresh water.

FACT CAT FACT

Elephants drink up to 200 litres of fresh water every day. That's more water than you could fit in a bathtub!

SOLIDS, LIQUIDS AND GASES

Water can have three **states**: **solid**, **liquid** or **gas**. The state of water depends on its **temperature**. When water **freezes**, it becomes a solid. When water reaches **boiling point**, it becomes a gas.

This is what water looks like as a solid, a liquid and a gas.

Water freezes at zero degrees Celsius. At what temperature does water boil?

As a liquid, water can flow into the shape of any container you put it in. **Water vapour** (a gas) can move through the air. You can see water vapour rising up from a kettle after it boils.

This iceberg is made of ice (solid water). Solid water is hard. When ice is heated, it **melts** into liquid water.

THE WATER CYCLE

Water is always moving from the surface of Earth to the **atmosphere** and back again. This movement is called the water cycle.

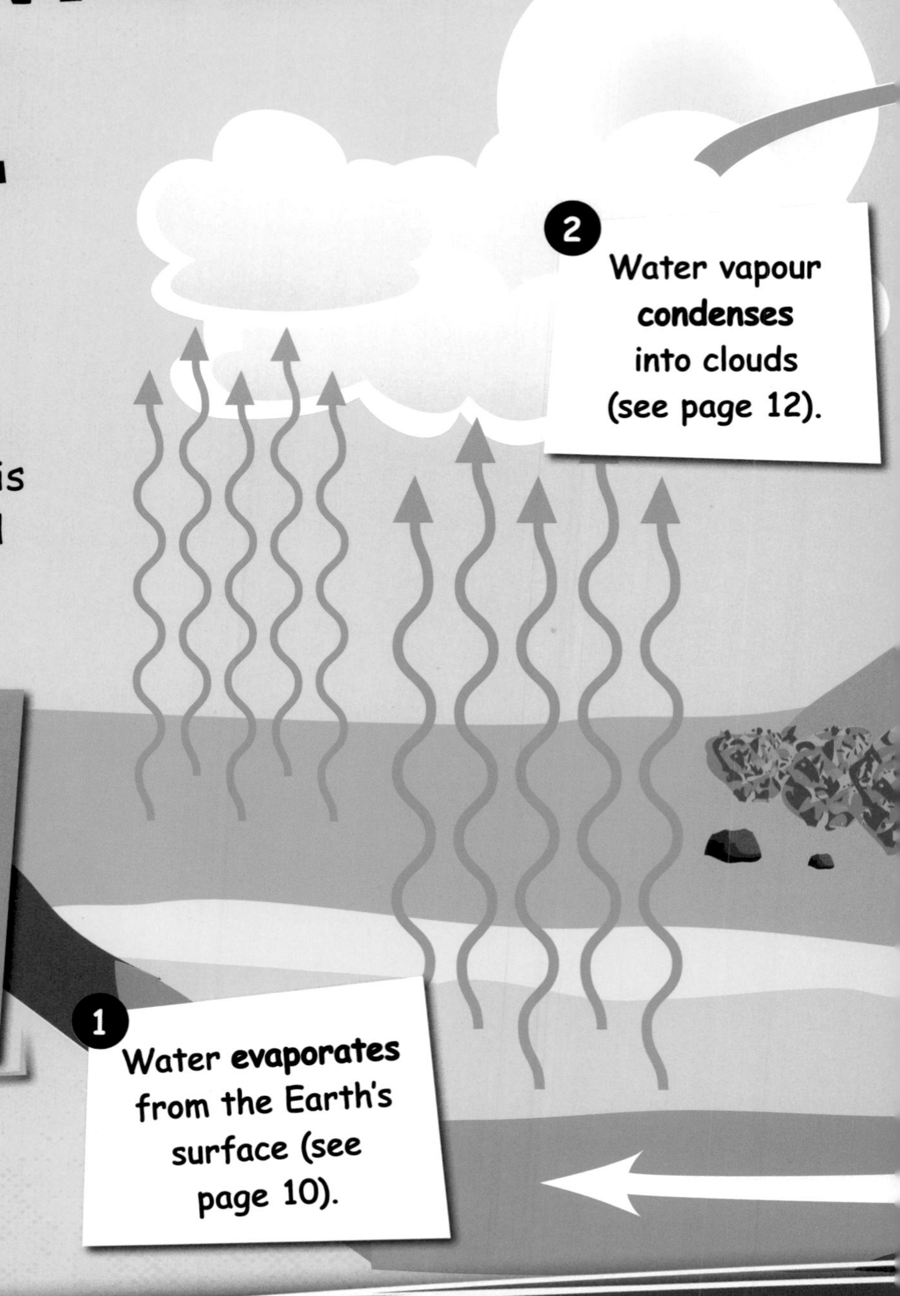

This drawing shows the stages of the water cycle. Where can you see solid water in this drawing?

1 Water **evaporates** from the Earth's surface (see page 10).

2 Water vapour **condenses** into clouds (see page 12).

3
Water falls back to the Earth's surface (see pages 14 and 16).
FACT CAT FACT
No new water is made or lost in the water cycle. This means that the water on Earth today is the same water that was around during the time of the dinosaurs!
4
Water flows back into the sea (see page 18).

EVAPORATION

At the beginning of the water cycle, heat from the Sun warms up water on the Earth's surface. When liquid water is heated, it evaporates and turns into water vapour.

Water evaporates from the surface of lakes, rivers and puddles.

Sometimes you can see evaporation taking place. If it rains on a hot day, the puddles of water on the pavement quickly disappear. This is because the liquid water has evaporated into water vapour.

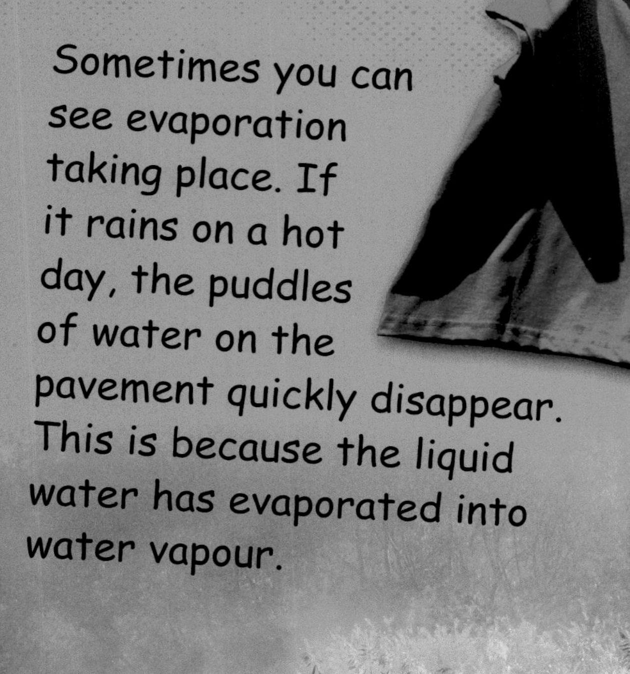

Wet washing dries when water evaporates from it. Why do clothes dry faster when they are hung outside on a sunny warm day?

FACT CAT FACT

When salt water evaporates, the salt it contains is left behind as a solid.

CONDENSATION

As water vapour is a gas, it rises up into the atmosphere. It is colder in the sky than on the ground. As the water vapour cools, it turns into clouds. This is called condensation.

Clouds are formed high in the sky.

Clouds are much heavier than they look. A small cloud can weigh as much as two elephants!

We can often see condensation on windows. The water vapour in the air cools down when it hits the cold glass. This creates small drops of water on the window.

This window is covered in condensation. Why do you often see condensation on the outside of a glass of cold drink?

RAIN AND SNOW

As the water vapour in clouds cools down even more, it turns into liquid water. This water falls to Earth as rain.

This girl is us umbrella to kee dry in the

Each drop of water spends about ten days in the sky before falling to the ground.

When the weather is really cold, rain freezes. It falls to the ground as snow or **hail**.

This girl has
a snowman
fallen snow
happen to
when the
gets

UNDERGROUND WATER

When rainwater reaches the ground, it soaks into soil on the Earth's surface. Plant roots take in some of the water, but most of it travels deep underground.

When rainwater moves through holes in the ground, it makes underground lakes.

Sometimes, underground water flows back above ground to form a **spring**. Springs are often the **sources** of rivers.

FACT CAT FACT

A river can have more than one source.

As water flows out of this spring, it forms a small waterfall. What is the tallest waterfall in the world?

RIVERS

Some of the water that falls as rain runs across the surface of the Earth. It flows downhill until it reaches a river.

Rivers get wider as more water flows into them. What is the name of the widest river in the world?

Eventually, all rivers flow into the sea. The water that reaches the sea will soon evaporate into water vapour and the water cycle will start again.

The place where a river meets the sea is called an **estuary**.

The water in an estuary is a mixture of salt water and fresh water.

CONTROLLING WATER

Humans can control the flow of water. We build large strong walls, called **dams**, to control how much water flows down a river. Dams stop rivers from **flooding** after lots of rain.

This dam is on the Orange River in South Africa. Find out the names of some other dams around the world.

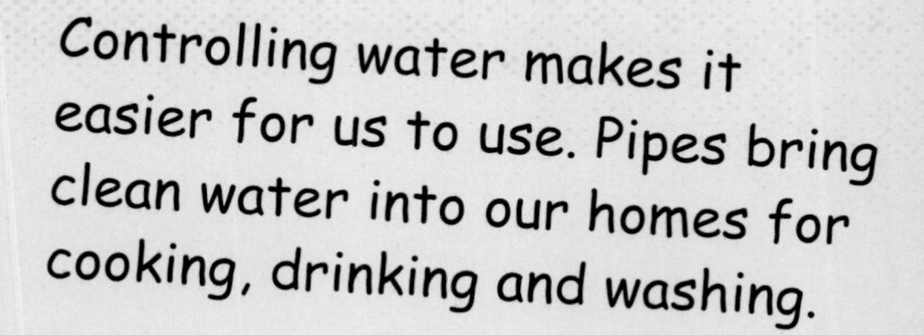

Controlling water makes it easier for us to use. Pipes bring clean water into our homes for cooking, drinking and washing.

Some people in the world don't have water pipes in their homes. These people in Ethiopia, Africa, are collecting water from a nearby lake.

FACT CAT FACT

Around one billion people around the world don't have clean water pipes in their homes or in their villages. They have to drink dirty water from lakes and rivers, which often makes them sick.

Try to answer the questions below. Look back through the book to help you. Check your answers on page 24.

1 **Most water on Earth is salt water. True or not true?**

a) true

b) not true

2 **Which state is ice?**

a) solid

b) liquid

c) gas

3 **Water evaporates when it cools down. True or not true?**

a) true

b) not true

4 **What is condensation?**

a) when ice warms up and melts into liquid water

b) when liquid water cools down and turns into ice

c) when water vapour cools down and turns into liquid water

5 **The source of a river is where it starts. True or not true?**

a) true

b) not true

GLOSSARY

atmosphere the air around the Earth

boiling point the temperature that a liquid boils at (100 °C for water)

condense to change from a gas to a liquid as it gets colder

dam a strong wall built across a river to control how much water passes through

estuary the place where a river flows into the sea

evaporate to change from a liquid to a gas as it gets warmer

flooding when a lot of water covers an area that is usually dry, often because a river has too much water in it

freeze when a liquid turns hard and solid because of the cold

fresh water water without salt in it, found in lakes and rivers

gas something that is neither a solid nor a liquid, such as air

liquid something that is neither a solid nor a gas, such as milk

living thing something that is alive, such as a plant or an animal

melt to turn from a solid to a liquid as it gets warmer

salt water water with salt in it, found in seas and oceans

solid something that is neither a liquid nor a gas, such as a table

source the starting point of a river

spring a place where underground water flows out of the ground

state the state of an object can be solid, liquid or gas.

surface the top part of something

survive to stay alive

temperature how hot or cold something is

water vapour water as a gas

INDEX

animals 4
atmosphere 8,12

clouds 12, 14
condensation 8, 12–13

dams 20

estuary 19
evaporation 8, 10–11, 19

floods 20
fresh water 4, 5, 19

gases (vapour) 6–7, 10, 11, 12, 13, 14, 19

hail 15

ice 7

lakes 10, 16, 21
liquids 6–7, 10, 11, 14

plants 5, 16
ponds 4
puddles 10, 11

rain 11, 14–15, 16, 20
rivers 4, 5, 10, 17, 18–19, 20, 21

salt water 5, 11, 19
seas and oceans 5, 9, 19
snow 14–15
solids 6–7, 8
source (of river) 17

water cycle 8–9, 10, 19
waterfall 17

ANSWERS

Pages 4–21

Page 5: In lakes

Page 6: 100 °C

Page 8: On top of the mountains there is ice (solid water).

Page 11: Because there is more heat from the Sun on a sunny day.

Page 13: Because the water vapour in the air condenses when it hits the cold glass of drink.

Page 15: It will melt and turn into a liquid.

Page 17: Angel Falls

Page 18: The Amazon River

Page 20: Some dams around the world include the Hoover Dam, the Aswan Dam and the Three Gorges Dam.

Quiz answers

1 true

2 a - solid

3 not true – it evaporates when it heats up.

4 c - when water vapour cools down and turns into liquid water.

5 true